W9-CHU-095

What is Weather?

Sunshine

Miranda Ashwell and Andy Owen

Heinemann Library
Des Plaines, Illinois

Published by Heinemann Library,
an imprint of Reed Educational & Professional Publishing,
1350 East Touhy Avenue, Suite 240 West
Des Plaines, IL 60018

Printed and bound in Hong Kong
Designed by David Oakley
Illustrations by Jeff Edwards

03 02 01 00 99
10 9 8 7 6 5 4 3 2 1

Library of Congress Cataloging-in-Publication Data

Sunshine / Andy Owen and Miranda Ashwell.
p. cm. – (What is weather?)
Summary: Briefly describes the sun and topics related to sunlight, including day and night, shadows, heat and cold, and the effects of sunshine on plans and animals.
ISBN 1-57572-791-9
Sunshine–Juvenile literature. [1. Sun. 2. Sunshine.]
I. Ashwell, Miranda, 1947- . II. Title. III. Series: Owen, Andy,
1961- What is weather?
QC911.2.084 1999
551.5'271–dc21 98-42815
CIP
AC

Acknowledgments
The author and publishers are grateful to the following for permission to reproduce copyright material: Bruce Coleman Limited/Atlantide, p. 26; Dr. S. Coyne, p. 20; H-P Merten, p. 24; Robert Harding Picture Library, pp. 7, 19; Robert Francis, p.14; R. Oulds, p. 15; P. Robinson, p. 27; Image Bank:/I. Royd, p. 28; Oxford Scientific Films/D. Cox, p. 17; T. Jackson, p. 23; S. Osolinski, p. 11; M. Pitts, p. 18; F. Polking/Okapia, p. 5; Pictor International, p. 6; Planet Earth Pictures/J. Lythgoe, p. 10; J. MacKinnon, p. 11; Science Photo Library/ NASA, p. 10; Still Pictures/M. Gunther, p. 22; H. Klein, p. 21; NASA, p. 4; T. Raupach, p. 25; Telegraph Colour Library/C. Ladd p. 28; F. O'Brien, p. 27; Tony Stone Images/T. Flach p. 12.

Cover photograph: Bruce Coleman Limited/F. Labhardt.

Some words are shown in bold, **like this**. You can find out what they mean by looking in the glossary.

Contents

What Is the Sun?

The sun is a star. A star is a burning ball of hot **gas**. The burning gas makes heat and light.

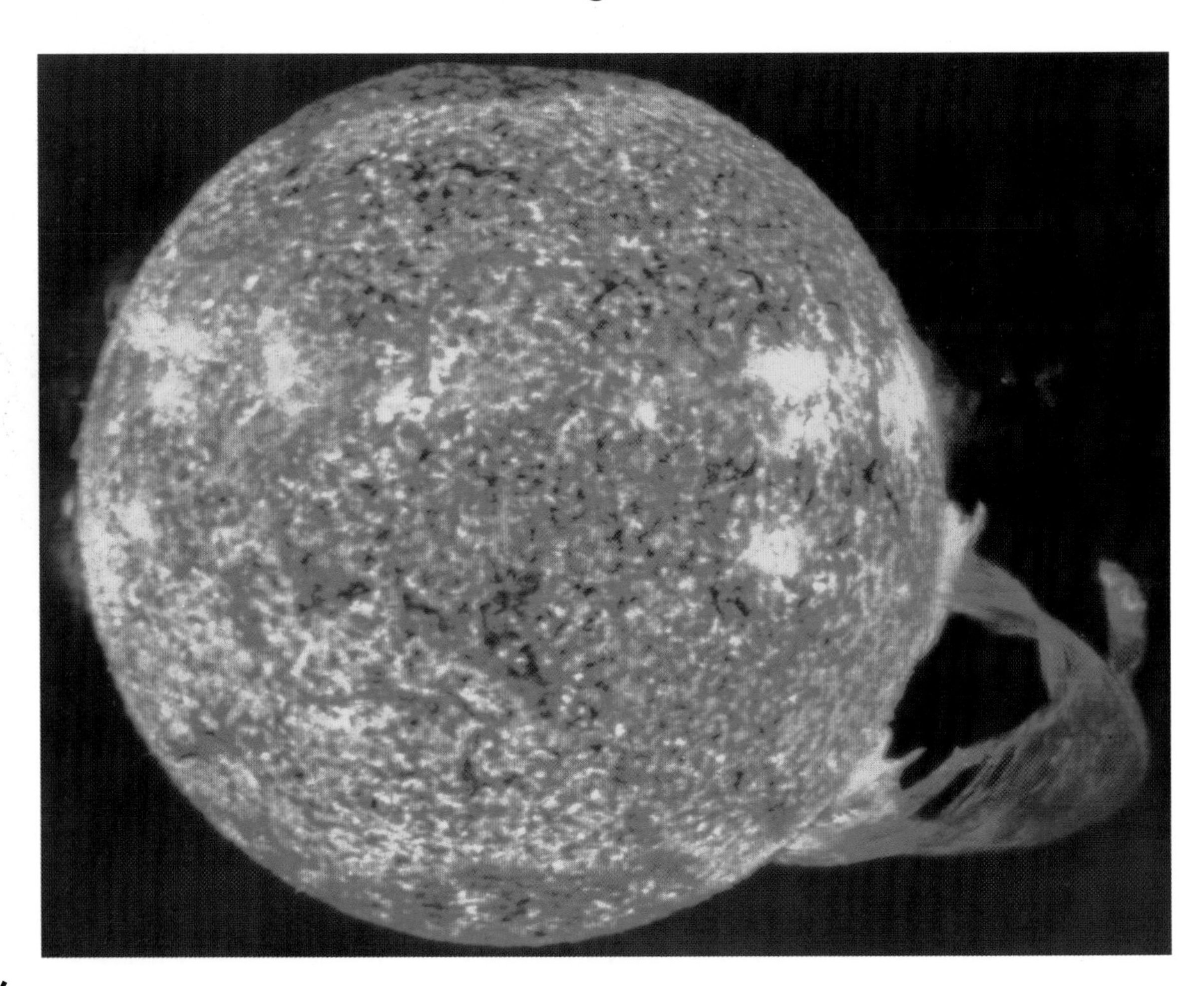

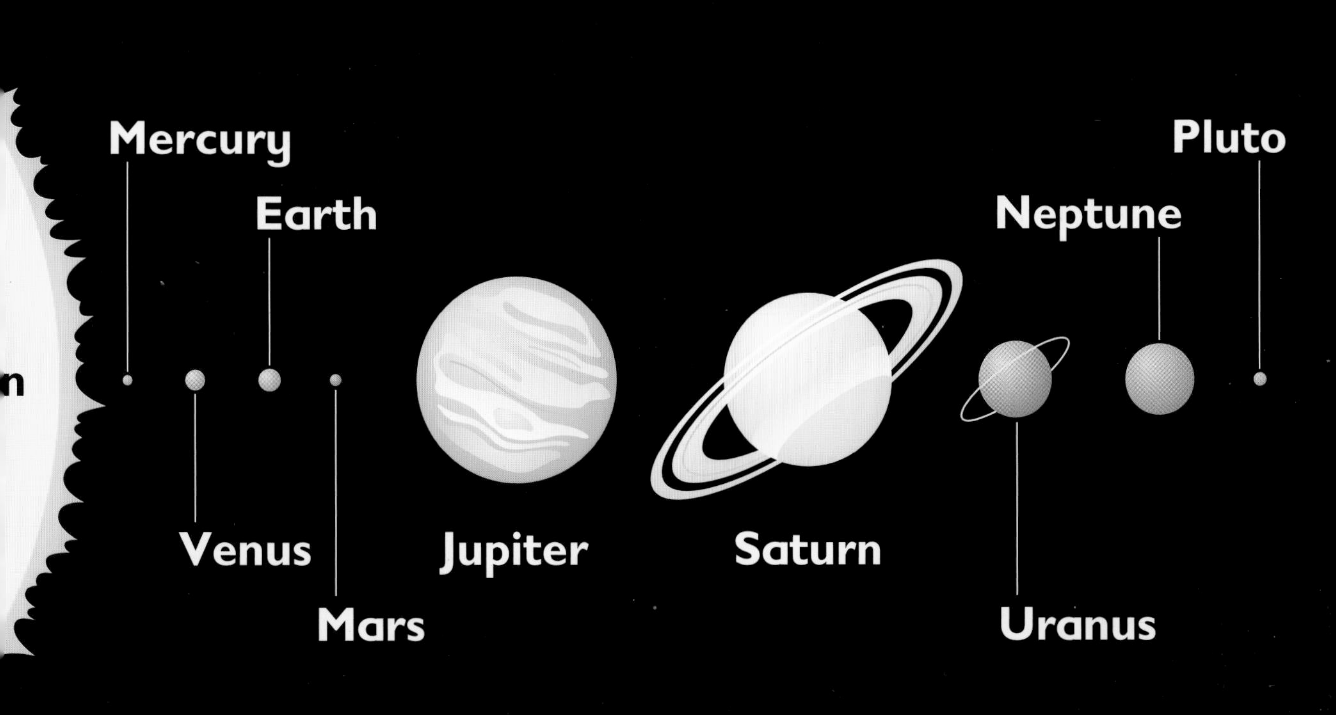

The sun is very many times larger than the earth. The earth is a **planet**. It is 93 million miles (150 million kilometers) away from the sun.

Heat from the Sun

Energy from the sun travels through space. It reaches the earth and warms the ground.

Air is warmed by the hot ground.
The warm air rises. Warm, rising air
pushed these clouds high into the sky.

Day and Night

The earth has day and night because it spins. On the sunny side of the earth it is daytime. On the dark side of the earth it is nighttime.

This is the sun rising in the morning sky. The sun seems to move across the sky and go down in the evening. But really the sun stays still while the earth is turning.

Seasons

Sometimes the part of the earth where you live is closer to the sun. The weather is warmer. This is called summer. The weather is colder in winter.

In countries near the **equator**, the sun is always nearly overhead. The weather is warm for the whole year.

Light and Dark

The weather is warmer during the day than at night. This is because the sun is shining on the other side of the earth at night.

During the summer, the sun shines on our part of the earth for longer each day. Nights are shorter and the weather is warmer.

Shadows

A **shadow** is made when light from the sun is blocked. Shadows are longest when the sun is low in the sky. They are shortest when the sun is high in the sky.

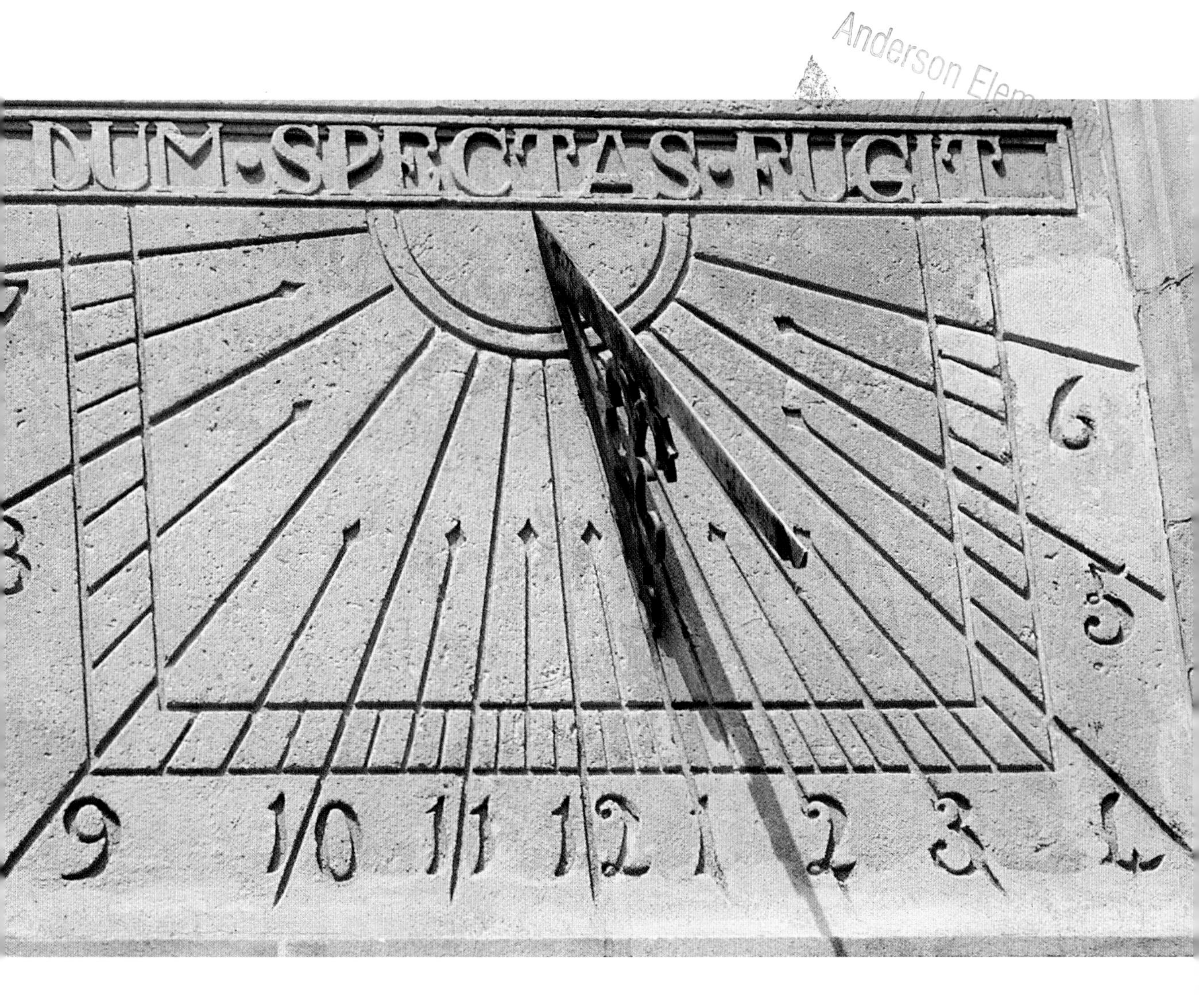

People can use shadows to tell the time. The metal finger on this **sundial** makes a shadow. The shadow moves through the day.

Hot and Cold

Heat and light from the sun are strongest at the **equator**. This is the hottest part of the world.

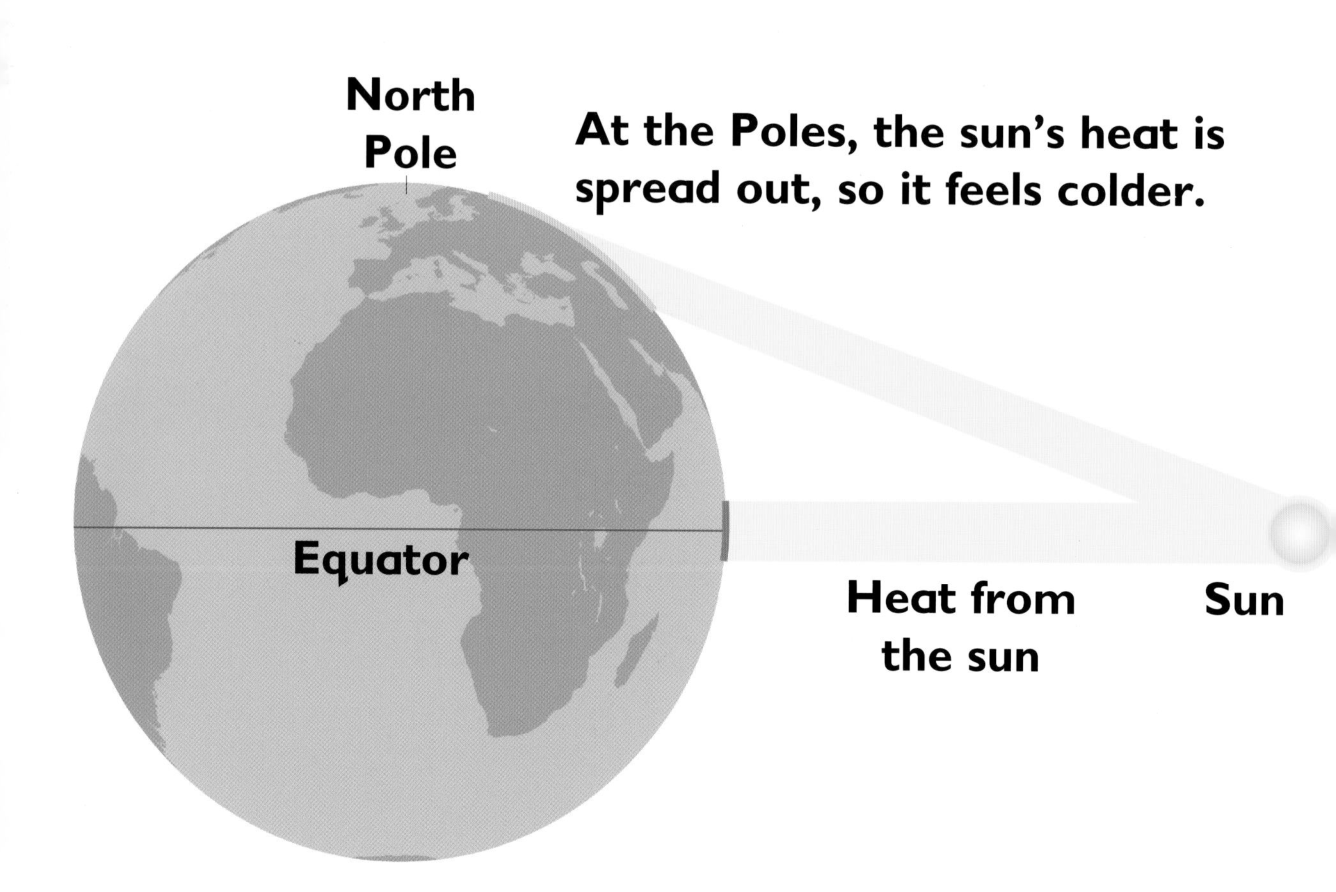

The sun never feels very warm at the North and South **Poles**. These are the coldest parts of the world. This is because the sun's heat is spread out at these places.

Plants and the Sun

Green plants use sunlight to make their food. The special green color in their leaves uses air and water to change sunlight into food.

Plants grow towards sunlight. The plants in this forest grow fast and tall in a race to reach the light. Plants with no light will die because they cannot make their food.

Special Plants

Some plants live in the **shadow** of taller plants. They have special leaves to make as much food as possible with only a little light.

Plants that live in very sunny places have special small, waxy leaves that protect the plant. They stop the plant from drying out in the heat of the sun.

Special Animals

Snakes and lizards use heat from the sun to warm their bodies. Warmth from the sun makes them more active.

This squirrel lives in a very hot, sunny place. She uses the **shadow** of her large tail to protect herself from the heat of the sun.

Sun Power

These **solar panels** catch energy from the sun. They turn it into electricity, which people use to heat and light their homes.

Solar panels make the energy for these electric cars. Most cars use gasoline, which makes the air dirty. The **solar energy** used by these cars is clean.

Enjoying the Sun

We enjoy resting or playing in sunshine. Many people go on vacation to warm, sunny places.

In hot countries, people often wear white clothes. White clothes keep people cooler than dark clothes.

Danger from the Sun

Energy from the sun can burn your skin and hurt your eyes. On sunny days, you protect your skin with sun cream. You should also wear a hat and sunglasses.

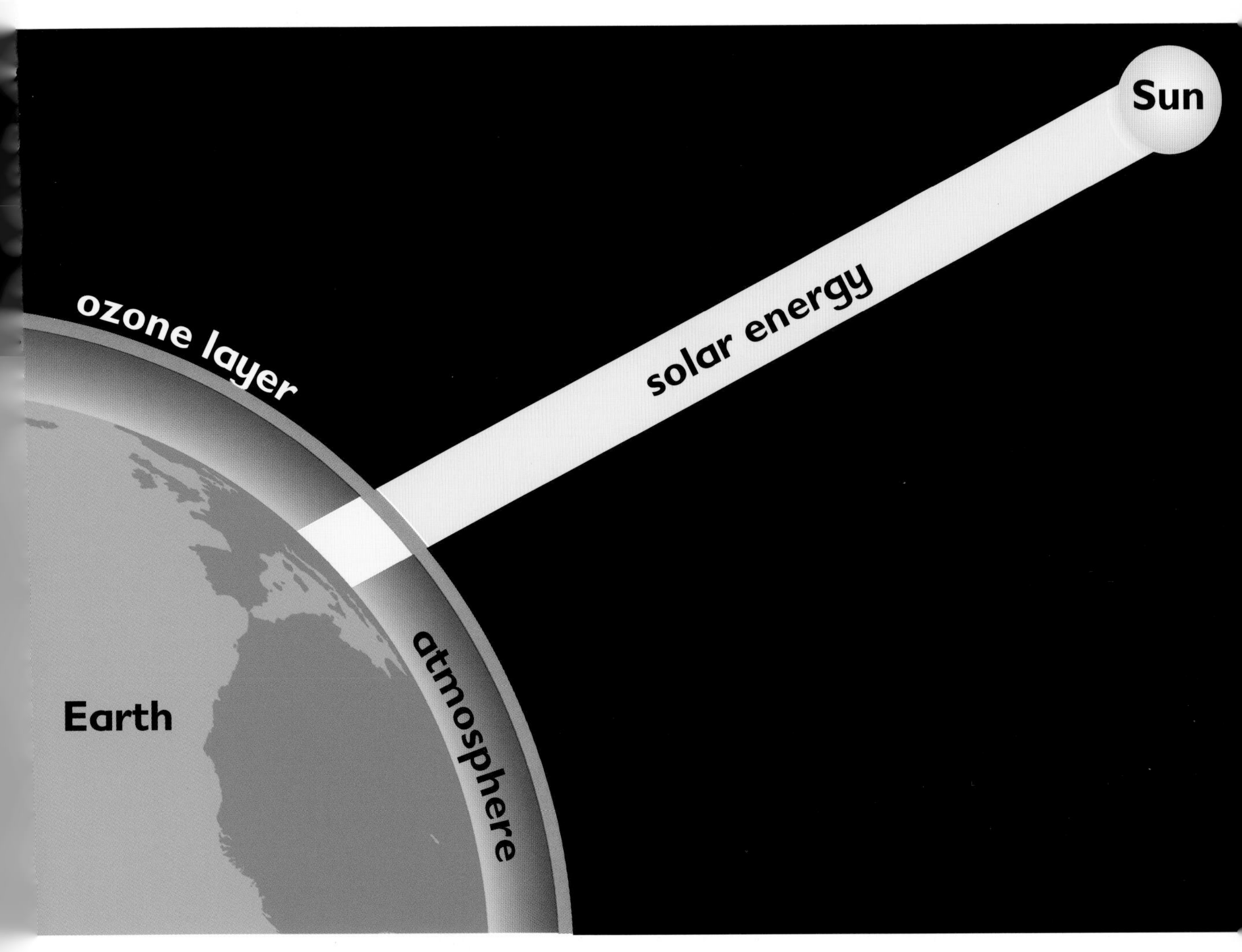

There is a layer of **gases** 15 miles (24 kilometers) above the earth. These gases block out much of the harmful **solar energy**. This layer is called the **ozone layer**.

It's Amazing!

The sun is 5,000 million years old. People think that the sun will burn for another 5,000 million years.

The sunniest place in the world is the Sahara **Desert**. The sun shines for 4,300 hours a year, or about 12 hours every day.

In the 1980s, scientists found a hole in the **ozone layer**. It is now even more important that you protect yourself from the sun's harmful energy.

Glossary

desert place where there is very little rain

equator pretend line around the middle of the earth, splitting the earth into two halves

gas form of matter that does not have a shape

ozone layer layer of gas around the earth that protects it from the sun's harmful energy

planet large object that circles around a star

poles the North and South Poles are the two points furthest from the equator

shadow dark shape made when something blocks light from the sun

solar energy heat and light from the sun

solar panel device that turns the sun's energy into energy we can use

sundial type of clock that tells the time with shadows

More Books to Read

Branley, Franklyn M. *Sunshine Makes the Seasons.* New York: HarperCollins Children's Books, 1985.

Palazzo, Janet. *Our Friend the Sun*. Mahwah, NJ: Troll Communications, 1997.

Saunders-Smith, Gail. *Sunshine*. Danbury, Conn: Children's Press, 1998.

Index